Tales in the Night Sky

A Gentle Introduction to Star Gazing

Tales in the Night Sky

A Gentle Introduction to Star Gazing

Rob Drew

Illustrated by Elaine Drew

TaeranArts
Pleasanton CA

INDIANA STATE UNIVERSITY

Copyright © 2011 by Rob Drew. All rights reserved. No part of this publication may be reproduced, stored in a retrieval system or transmitted in any form or by any means including electronic, mechanical, photocopying, recording, or otherwise without the prior written permission of the copyright holder, except for brief quotations used in a review.

Cover and interior design by Elaine Drew

Illustrations copyright © 2011 by Elaine Drew

Published by TaeranArts Publishing, Pleasanton, CA

ISBN 978-0-9833236-9-3

For Tasha and Tamsen

CONTENTS

INTRODUCTION	ix
THE AUTUMN SKY	1
What to look for in the sky	2
Cassia's Crown	4
The Dark Tower of Cepheus	7
Andromeda's Chain	9
The Great Square of Pegasus	10
Aries the Skinny Sheep	11
The Shield of Perseus	12
The Pleiades	14
THE WINTER SKY	17
What to look for in the sky	18
Orion's Belt	20
Taurus the Bull	22
The Chariot Wheel	23
Major the Big Dog	24
Minor the Little Dog	25
The Twins	26
THE SPRING SKY	27
What to look for in the sky	28
Leo the Lion	30
The Chariot, the Plow, the Dipper	33
The Boots	34
Virgo's Spike of Wheat	36
THE SUMMER SKY	37
Early Summer: Hercules and the Golden Apples	38
What to look for in the sky	38
The Golden Apples	40
Draco the Dragon	41
Hercules	42

MIDSUMMER: THE THREE KINGS	45
What to look for in the sky	46
The Staff of Solon	48
The Scorpion	49
The Persian Archer	50
The Python of Delphi	52
LATE SUMMER: THE SUMMER TRIANGLE: THE LYRE, THE SWAN, AND THE EAGLE	53
What to look for in the sky	54
The Lyre of Orpheus	55
The Swan Hang Glider	56
The Eagle Hang Glider	58
STARS AND PLANETS	61
Turning on the Stars	63
List of Bright Stars	64
Constellations of the Zodiac	66
GLOSSARY	67
INDEX	69

STAR MAPS

Autumn Star Map	3
Open Star Clusters and the Andromeda Galaxy	15
Winter Sky Map	19
Spring Sky Map	29
Early Summer Star Map	39
Midsummer Summer Sky Map	47
Late Summer Star Map	54

INTRODUCTION

This is a picture book of stories about the night sky. Traditionally star stories are borrowed from ancient Greek myths. Storytellers would group stars together into "connect the dots" pictures to fit their tales. These groups of stars are called constellations from the Latin word meaning "groups of stars." We can't usually see as many stars as the Greeks did, because of our city lights, so it's difficult to see the old constellation pictures. Some new pictures would certainly help, and that's what this book is about.

Our sky maps feature only the stars that you're most likely to see from a modern suburban neighborhood. We connect these bright dots to form new star pictures, which astronomers call asterisms. Simple asterisms can make constellations easier to find. For example, instead of trying to find the entire sword-wielding hero Perseus, we will look only for the stars that form his famous shield. Two famous asterisms you may already know are the Big Dipper and Orion's Belt. These star pictures are parts of the larger constellations Ursa Major and Orion the Hunter.

Each of our star pictures represents someone or something in one of our stories. The book is organized by the seasons of the year, so that all the characters in one story will be near each other in the sky. The stories will help you remember which constellations to look for depending on the time of year. For example, since Leo the Lion is in a spring story, there's no point in looking for him in the fall. Since Orion and Taurus the Bull are in a story together, when you find one you know the other must be nearby.

Autumn, winter, and spring each have one story. Summer has three stories because there are lots of summer stars, and it's a nice time to be outdoors. Each story has several constellations in it, so by the end of the book you should recognize 26 constellations. You'll also know the names of over 20 individual stars.

Sometimes there is an interesting fact about the stars in a constellation. We'll point this out so you can impress your friends. Also, some constellations have special objects like star clusters, glowing nebulae or a distant galaxy that you might be able to see in binoculars if you have some.

We hope these pictures and stories will make it easier for you to locate and identify constellations and the brightest stars in the night sky. The next time you are outside on a beautiful evening, maybe you'll remember one of these stories. Then you can share it with your family or your friends while you point out the stars that represent the pictures.

THE AUTUMN SKY

Our autumn story is found in the stars of late September, October, and November. This is after the equinox, when day and night are each about twelve hours long. Six constellations, a star cluster, and a neighboring galaxy all play a part. The characters are the queen and king of Jaffa, a damsel in distress, a big rock, a little sheep, some mischievous breezes, and a handsome young hero. It is the story of Cassia and Cepheus, Andromeda and Perseus. You'll find their star shapes high overhead, spreading from west to east.

Once upon a time, Cassia and Cepheus ruled a small, peaceful kingdom called Jaffa on the sea coast near Egypt. They had a lovely daughter named Andromeda. In those days, queens were more important than kings, so Queen Cassia was the ruler. She cared for the prosperity and well-being of her people, and was loved by them. Cepheus ran the family business, organizing trade between merchants from Crete and wealthy Syrian and Egyptian families. In a beautiful garden near the shore, Cepheus built a tall tower, where he could study the stars and watch for the merchants' ships. On stormy nights he would build a beacon fire to guide the ships to shore.

When Andromeda was old enough, her parents arranged for her to marry Aegenor, a Syrian king from the town of Tyre. But Andromeda wanted to choose her own husband. She refused to marry Aegenor, because he was too old. Instead she chained herself to a giant rock by the sea, saying she would rather be eaten by a sea monster. This rock was a famous spot, because it had the outline of Pegasus, a flying horse, carved on it. There she sat, with only one little sheep for company.

While Andromeda was sitting on her rock, waiting to be eaten, a marauding young hero named Perseus happened along. He had recently defeated a demon named the Gorgon, and had cut off its head. While on his way home his ship had been blown towards Jaffa by the mischievous Pleiades, daughters of the wind. Perseus noticed Andromeda sitting by the sea. She told him her sad story, and Perseus promised to save her. He spotted a nearby sea monster and drove it away. Then he returned hoping to claim Andromeda as his bride. Fortunately for our story, she thought Perseus was very brave and handsome. She chose to marry him, and they all lived happily ever after.

What to look for in the sky

The great square of Pegasus is formed by four stars to the east of the Summer Triangle. The summer star Deneb in the Swan's tail points toward the tower of Cepheus. The tower is "dark" because the stars are very faint. East from the tower, look for Cassia's Crown, shaped like a big letter **W**. Andromeda's chain is the string of stars leading eastward from the great square. At the end of the chain you'll find Perseus' shield, reflecting the dreaded Gorgon's eye. You can also locate the shield by finding the Pleiades star cluster and then looking a little toward the north.

- *Cassia's Crown*
- *The Dark Tower of Cepheus*
- *The Great Square of Pegasus*
- *Andromeda's Chain and her Mirror*
- *A very skinny sheep named Aries*
- *The Shield of Perseus and Algol, the Gorgon's eye*
- *The Pleiades star cluster*

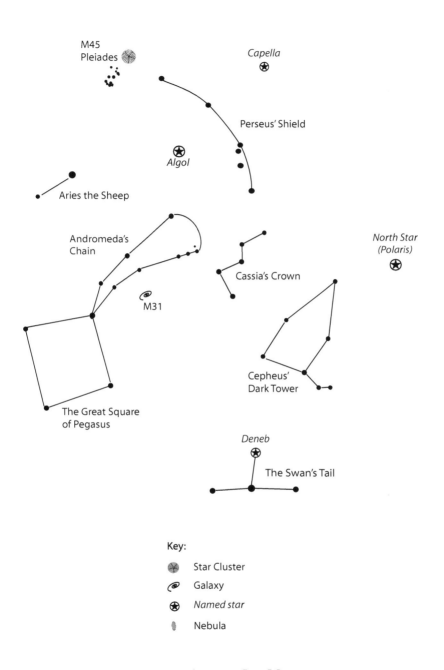

Autumn Star Map
Face west and look up

Cassia's Crown
CONSTELLATION: CASSIOPEIA THE QUEEN

In the beginning of time, many places were ruled by women. Jaffa was ruled by a queen, and her name was Cassiopeia, Cassia for short. She was named for the cinnamon tree. Perhaps cinnamon was the color of her hair, or perhaps she liked cinnamon rolls for breakfast. Nobody knows for sure, but her reign was prosperous and peaceful. There were no walls around her palace. Instead it was surrounded by a garden with fountains and had a beautiful courtyard lined with palm trees.

Cassia was determined to be a good queen, and she chose Cepheus as her king consort because he was an honest and thoughtful man. Since her country was prosperous, many wandering people began to move there. Soon there were more newcomers than citizens, with many different customs and languages. The Nereids were especially difficult, demanding a new queen be chosen every year in a beauty contest. Cassia thanked them for their suggestion, but explained that being a queen is a serious job. Aided by Cepheus, Cassia endured and continued to rule well for many years.

The star picture for Cassia is her crown. It's pointy like a letter **M** or **W** with one bent corner. Look for it high overhead in the fall and toward the northwest in the winter.

In binoculars

Cassia's Crown is in the middle of an arm of our galaxy, the Milky Way, and is filled with star clusters. These are small groups of stars, maybe a hundred or so, that formed from the same cloud of gas and dust. Try looking toward the crown on a dark, clear night, and see if you can find some of these star clusters.

Queen Cassiopeia: You can see Cassia's Crown in the autumn sky.

Cepheus looks out from his Dark Tower.

The Dark Tower of Cepheus
CONSTELLATION: CEPHEUS THE KING

Since Jaffa was ruled by a queen, King Cepheus didn't have much to do. He loved Cassia, and liked to keep busy with his garden. He collected animals and plants from far-off places, brought to him by traders from Syria, Egypt, Ethiopia, and Crete. His favorite animal was his giraffe. On holidays, he would invite the people of Jaffa to bring their children to enjoy the gardens.

Cepheus was very curious about the world, and brought wise men to the court to study with him. He also made a hobby of measuring things. How tall was a distant tree or hill? How far from his tower was a certain ship? He had some men count their footsteps when walking to nearby towns. He even wondered how to measure the size of the moon and the distance to the sun and the stars.

The star picture for Cepheus is his castle tower. It looks like a house with a pointed roof. The stars of the castle are very faint, so look for it just west of Cassia, with the point of the roof near Polaris, the North Star. The star Deneb at the tip of the Swan's tail also points to Cepheus' Tower.

Impress your friends

King Cepheus would be very pleased that a system of measuring astronomical distances uses a kind of star discovered in his constellation. *Delta Cephei* is a type of star called a **Cepheid variable**. Its brightness varies over a period of five days. In 1912 Henrietta Leavitt, a Harvard astronomer, found that the true brightness of a Cepheid variable is related to its cycle: the longer the cycle the brighter the star. As stars get farther away from us, they look dimmer. So if we know how bright a star really is, and measure how bright it looks, we can calculate its distance. With Cepheid variables, the cycle time tells the true brightness. In 1924, Edwin Hubble used Cepheid stars to measure the distance to the Andromeda galaxy, 2 million light years away.

Andromeda's Chain

Andromeda's Chain
CONSTELLATION: ANDROMEDA THE PRINCESS

Andromeda was a beautiful, headstrong girl. Born a princess, she wanted to run her own life. When her mother arranged a marriage for her to old king Aegenor, she rebelled. She chained herself to a rock saying, "I'd rather be eaten by a sea monster." Fortunately, a charming hero named Perseus came along to rescue her and win her heart. As a wedding gift Perseus made her a necklace from a piece of her chain and some precious gems from the Gorgon's treasure hoard. She gave Perseus a sail for his boat, decorated with an embroidery of Pegasus the flying horse.

Before she sailed away with Perseus, Andromeda's father gave her a special mirror, in which she could always see her home. Looking at this mirror in the sky, we see what looks like a reflection of our own Milky Way Galaxy. In fact it is our neighbor galaxy, named after Andromeda.

The star picture for Andromeda is her chain, attached to the Square of Pegasus. Look also for her mirror, the Andromeda Galaxy, lying next to the chain.

In binoculars

The Andromeda Galaxy (M31) is easy to find in binoculars, and looks like a big fuzzy smudge. To see the Andromeda galaxy's spiral shape, you would need a fairly big telescope. When you find the smudge, you are seeing the light from billions of stars located 2 million light years away, totally outside our own galaxy.

Impress your friends

In 1999 three planets were discovered orbiting one of the stars in the constellation Andromeda. This was the first system of multiple planets found outside our own solar system.

The Great Square of Pegasus
Constellation: Pegasus the Flying Horse

No one has seen a flying horse, at least not recently, and you won't see a picture of one by connecting the dots in this constellation. But people back then liked stories about wonderful creatures and their adventures. Stories about Pegasus were so popular that artists from Jaffa carved a giant image of the flying horse on the surface of a great square rock outside their city. Every year the citizens held a festival on the shore near the rock. People brought picnic lunches to enjoy, and the children climbed the rock. This is the rock that Andromeda chained herself to when she was angry with her parents.

Our star picture is called the Great Square of Pegasus. Look for four stars that form a huge square in the sky, almost straight overhead. The stars are a bit faint, but the square can usually be seen because there aren't many stars in this part of the sky. You'll have to imagine your own flying horse.

Aries the Skinny Sheep
Constellation: Aries the Ram

It got a bit lonesome for Andromeda sitting out on her rock. Her only companion was a very skinny sheep named Aries. He spent the day munching grass. Aries didn't have much to say, but he was a good listener.

The star picture for Aries has only two stars. They are to the south of Andromeda, lying beside her chain.

> ## *Impress your friends*
>
> Aries is a constellation of the zodiac, twelve constellations that lie along the path that the sun, moon, and planets follow. This is the plane of our disk-like solar system, and is called the ecliptic. You can tell that our solar system is tipped relative to our galaxy, because the zodiac doesn't line up with the Milky Way.

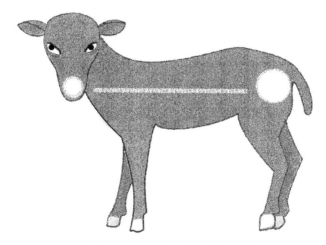

Aries, the Skinny Two-Star Sheep

The Shield of Perseus
Constellation: Perseus the Hero

Perseus was an adventurous fellow. He wanted to do something brave to impress the king of Seriphos. He boasted that he would kill the king's enemy, a cruel demon called the Gorgon. This wasn't an easy task, because the Gorgon lived far away. She also had the power to turn people into stone if they looked in her eyes. Perseus armed himself with a sword and a shiny shield of polished bronze. After a long search, he found the Gorgon's castle. When Perseus went inside to find her, he used his shield as a mirror. By looking at her reflection rather than the Gorgon's eyes he was not turned to stone. For her part, the Gorgon was so surprised to see herself in this mirror that Perseus was able to cut off her head and escape. He was on his way home when he stopped to rescue Andromeda, and you know the rest of that story.

The star picture for Perseus is his shield, lying above (north of) the Pleiades star cluster. Look for the Gorgon's reddish eye just to the west of the shield.

Impress your friends

The Gorgon's eye is a star called Algol. It is also called the *Demon Star*, because it seems to wink every 3 days for a few hours. How does it do that? Algol is really two stars, circling each other. The orbit of stars is in line with us, so when one star passes in front of the other some of the light is blocked. We call Algol an **eclipsing binary variable** star.

Perseus' Shield with Algol, the Eye of the Gorgon

The Pleiades
Star Cluster: The Seven Sisters

These stars represent seven sisters, daughters of Pleione, goddess of the breeze that fills the sails of ships. The Pleiades were sailing their own boat around the Mediterranean Sea. They saw that Andromeda was having problems, so they blew Perseus' ship in her direction.

The star picture for the Pleiades is a tiny dipper-shaped cluster of stars located east of Andromeda and just below Perseus' shield. Some people see six stars, some see more. With binoculars you can see a few dozen. There are about 500 stars in the cluster. This is an **open cluster**, a group of stars all born out of the same cloud of gas.

Impress your friends

Some night sky objects are identified by their *M* numbers. The M stands for Charles Messier, a French comet hunter. He made a catalog of interesting fuzzy things in the sky that were not comets, such as star clusters, galaxies, and glowing gas clouds called nebulae. For example, the Pleiades cluster is M45; the Andromeda galaxy is M31.

In binoculars

There are several open star clusters in Perseus, which can be found easily in binoculars. These are similar to the Pleiades, but look smaller because they are farther away. M34 is halfway between Algol and Andromeda. The famous Perseus Double Cluster is midway between Perseus and Cassia. These stars look like jewels, so we'll call them the Gorgon's treasure.

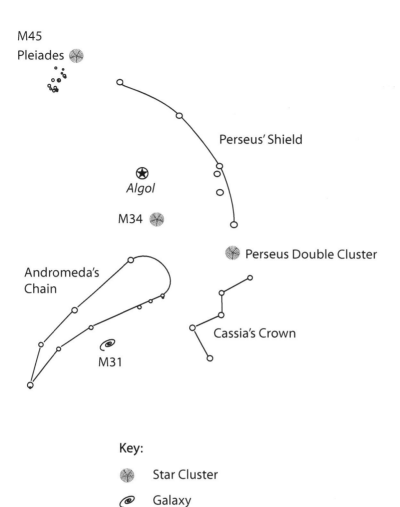

Open Star Clusters and the Andromeda Galaxy
Face west and look up

Orion, Taurus the Bull, and Auriga wearing his Capella

THE WINTER SKY

The winter sky is special because it holds half of the brightest stars, including the two closest to Earth. Our winter characters are found in the dark evenings of late December through March.

Once upon a time there was a giant hunter named Orion. He was on his way to visit his friends Castor and Pollux, the twin kings of Sparta. As he marched along he heard a loud bellow and a cry for help! Orion ran toward the sound and saw an enormous bull charging at a man who was hiding behind a chariot wheel. Orion noticed a brightly colored charioteer's cap peeking over the top of the wheel. There was no sign of any chariot, and the wheel was not a very good shield from the bull's horns.

Orion recognized the bull as Taurus, a ferocious animal with one red eye who did not like people traveling through his pasture. Bravely, the young hunter distracted the bull with his cape. Orion's big dog, Major, ran about barking at the bull. This was all too much for Taurus, and he retreated to the far corner of his pasture. The man with the cap thanked Orion for rescuing him, and said his name was Auriga. He was the chariot driver for Queen Demeter. A wheel had fallen off her chariot. Auriga had gone back to get it when he was attacked by the bull.

Orion looked around for his dogs. He found Major easily enough, but Minor, the little dog, had wandered off and was barking at something over the next hill. Orion saw Auriga safely on his way, and then continued his own journey to see the Twins.

What to look for in the sky

Sirius, the "dog star," is the brightest star in the sky, and forms the eye of Major the Big Dog. You'll find him low in the Southern sky. Trace upward to Orion, with bright stars at his foot and shoulder, and a beautiful three-star belt. Higher still is the bright star Capella, forming the Charioteer's cap. Next look for the horns and the bright red eye of Taurus the Bull. The horns are just below the chariot wheel; the eye is a little to the west. Looking east from the charioteer, you'll see two bright stars representing the Twins. Finally, there's one bright star between the Twins and Major. This is Procyon, which marks the eye of Minor the Little Dog.

- *Orion the Hunter and his belt*
- *Major the Big Dog with the very bright eye*
- *Auriga and his Chariot Wheel*
- *Taurus, the crazed Bull with a red eye*
- *The Twins Castor and Pollux*
- *Minor the Little Dog*

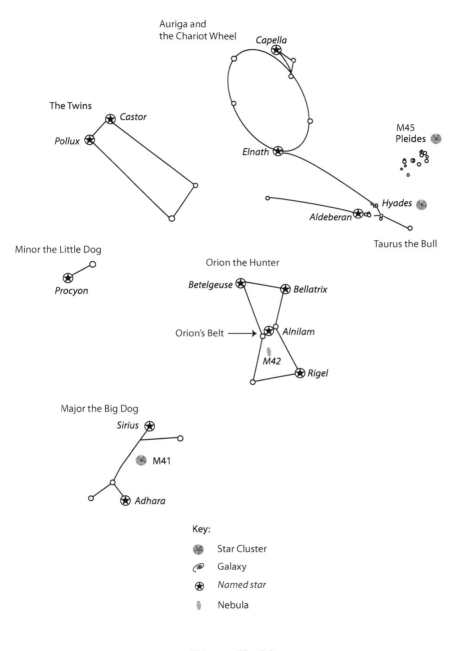

Winter Sky Map
Face south and look up

Orion's Belt
CONSTELLATION: ORION THE HUNTER

Orion is a very old constellation representing a hunter. He's so old there aren't many stories about him, but his starry belt is very famous. Perhaps he won it as a wrestling champion. Hanging from his belt is a tinder box. People in the old days carried fire-making materials with them, and sometimes smoldering embers rolled up in damp moss. In fact there are "sparks" visible in Orion's tinder box. These are bright, newly born stars.

The Milky Way galaxy has several spiral arms. Our sun is in the Orion Arm. When we look at Orion we are looking toward the outer edge of the galaxy. Several of Orion's stars are very far away but still look bright. That means they are much bigger and more luminous up close than either our sun or even Sirius, the closest bright star. Check the star table to learn more about Orion's brightest stars.

The star picture: Orion is shaped like a large hourglass, with stars at his knees and shoulders. He is easily recognized by his belt, three stars in a line.

Impress your friends

Compare the stars Rigel and Betelgeuse (pronounced "beetle juice"). Rigel, Orion's knee, is a hot, blue super-giant. Betelgeuse at his shoulder, is a cool, red super-giant, big enough to extend past Mars if it were our sun. Betelgeuse has a twin on the opposite side of the sky: the red super-giant Antares in the summer constellation of the Scorpion.

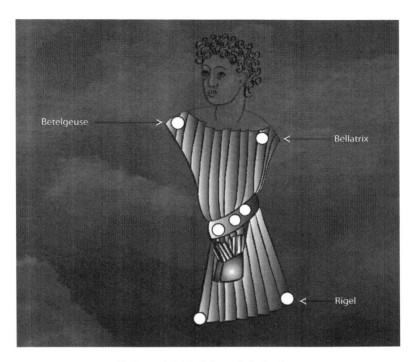

Orion with his belt and tinder box

In binoculars
Below Orion's Belt is the Great Orion Nebula, M42. This nebula is a cloud of dust and hydrogen gas where stars are being born. M42 is called an emission nebula. The new stars put out so much high-energy radiation that the nearby gas glows like a neon sign.

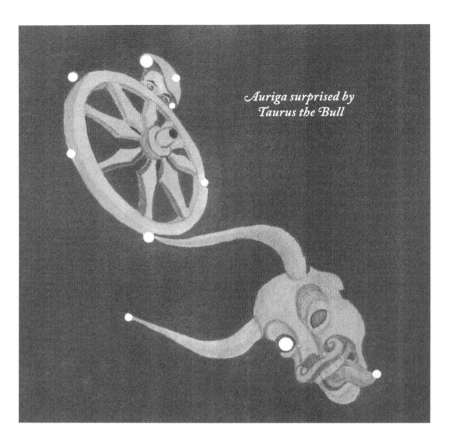
Auriga surprised by Taurus the Bull

Taurus the Bull

Taurus the red-eyed bull is another sign of the zodiac; it lies along the sun's path through the sky, called the ecliptic. Sometimes you'll notice the moon is in Taurus, and once in a while you might see a bright extra "star" there. This could be one of the planets: Jupiter, Saturn, Mars, or Venus, since they also follow the ecliptic.

The star picture of Taurus is a little disappointing. There are only three stars. Look for the reddish eye, marked by the star Aldebaran. The other two main stars are the tips of the bull's horns: Elnath, touching the Chariot Wheel, with the other below it.

The Chariot Wheel
Constellation Auriga the Chariot Driver

In the old days, people rode around in chariots instead of taxi cabs. If you're old enough to remember the TV show *Taxi*, you might recall one of the drivers named Rieger. That will help you remember our taxi driver's name, Auriga. Well, our taxi has a flat tire, or maybe the wheel fell off the chariot. Auriga was fixing it, when he was surprised by Taurus the Bull. You can just about see Auriga hiding behind the wheel. Look for the bright star on his cap, called Capella.

So now we've found the wheel and the driver, but where is the chariot? It will be along in a few weeks, high up in the spring sky.

The star picture is a roundish wheel, with the bright star Capella at the top. One of the bull's horns is touching the bottom of the wheel.

Impress your friends

Capella is actually a double star, two stars so close together that they look like a single star to us. These stars orbit each other about 3 times per year, and are closer to each other than Earth is to the Sun.

In binoculars

West of Taurus you can see the Pleiades, an open star cluster that we talked about in the autumn. The Hyades is also a star cluster, made of dim stars. They form the faint V-shaped face of the bull.

Major the Big Dog
CONSTELLATION: CANIS MAJOR

Some dogs are big and some are small. Major is truly a major dog. He always seems good-natured, romping through the winter sky scaring up rabbits and things for Orion to hunt. His bright eye is a star called Sirius, known as the Dog Star. It is so bright that some ancient people thought it contributed to the sun's heat when it was in the daytime summer sky.

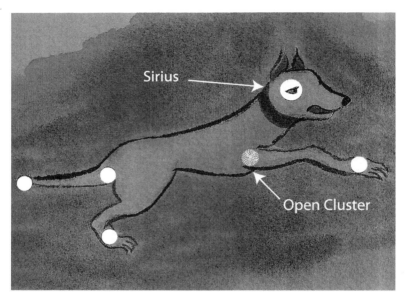

Major the Big Dog

Impress your friends

Sirius is both the brightest and the closest of our bright stars, about 8 light years away. Alpha Centauri is closer at 4 light years, but can be seen only in the southern hemisphere, for example Australia or South America.

In binoculars

In binoculars: If you look right about where the dog's heart would be, you might be able to see a gorgeous open star cluster, number M41 on the French Professor's list. However, if the moon is bright you probably won't see it.

Minor the Little Dog
Constellation: Canis Minor

Minor is a small dog, a two-star dog. He might be a sheep dog, because his two stars look a lot like Andromeda's pet sheep, Aries. The brighter star is called Procyon.

It looks as if Minor isn't paying any attention to the bullfight. In fact he's looking toward the east and barking with all his might. What do you suppose has him so excited? Well, you might make some noise, too, if you noticed Leo the Lion coming up over the horizon. But that's a story for the next chapter, about the Spring sky. When Minor's bright star Procyon is high in the west you can say it is announcing the coming of Spring.

The Twins
CONSTELLATION: GEMINI

Twins are kind of special even today, but in ancient times twins were considered very remarkable. The town of Sparta once had twin kings, named Castor and Pollux. Perhaps you didn't know that these boys had two sisters who were also twins. One was named Clytemnestra, who married a man named Agamemnon. Her twin sister was the famous Helen of Troy.

Of course brothers sometimes compete with each other. Well, in ancient times they say Castor was the brighter brother, but now Pollux is. No one is sure if Pollux is getting brighter or Castor is getting dumber . . . I mean dimmer.

Impress your friends

Appropriately for a twin, Castor is a double star. Each of those stars is also a double star, and a third double is nearby. So the "star" we see as Castor is actually a six-star system, with three sets of twins. Zeta Geminorum below Castor is a cepheid variable star.

The Twins are part of the Zodiac. If you see the moon in Taurus on one evening, you'll notice it has moved to the Twins a couple nights later.

In binoculars

To the east of the Twins, in the direction of Leo, is a very beautiful open cluster, called the Beehive Cluster or M44. It is not as bright as the Pleiades, so you'll need your binoculars. Slowly scan eastward between Pollux (one of the Twins) and Regulus (in Leo). It is about one-third of the way along between the two. You'll know it when you see it!

Note: See the Spring Star Map for the location of M44.

THE SPRING SKY

The arrival of Leo the Lion heralds the coming of spring. April, May, and June are the months to look for our next group of star pictures. Our spring sky story has only four constellations, but one is a very versatile actor, and plays three separate roles. Talk about fast costume changes!

This ancient story tells how people in Greece learned to grow wheat. There are several versions, but they all start with a lady looking for her lost daughter. Sometimes the lady is called a queen, sometimes she is called a goddess. Her name is Demeter, which means Mother Earth, and she traveled about in her own Chariot. This wonderful chariot could fly through the air, and was pulled by a giant lion named Leo. Demeter's daughter Persephone had gone missing, and her mother was very worried. She set out to look for Persephone, flying over the land of Greece, but her daughter was nowhere to be found. Finally, Demeter stopped to rest near a small village.

There she met a herdsman named Triptolemus who was tending his herd of cows. He was a little frightened by the lion, but he could see that the lady was tired and thirsty. He filled a long-handled Dipper with cool water from a nearby stream and offered it to her. Demeter was very grateful for his kindness. She thanked him, and then asked if perhaps he had any news of her daughter. Triptolemus didn't know Persephone, but he had seen a dark-haired girl carried off by a ne'er-do-well character named Pluto.

Demeter knew who Pluto was and determined to round him up for a very stern conversation. But first, she presented some gifts to Triptolemus in gratitude for his help solving the mystery. She showed him how to make a Plow, and use it to till a field for planting wheat. "With this plow," she told him, "one man can grow enough wheat to feed a whole village." Then she gave him a splendid pair of lizard-toed, cowhide Boots so everyone would know he was an important person and a favorite of Demeter. As she mounted her Chariot, she handed Triptolemus a precious Spike of Wheat, containing the seeds to plant for his first crop.

What to look for in the sky

First, find Leo, resting eastward in the sky behind the Twins. Then find the Chariot/Dipper/Plow, right above the lion. Follow the arc of stars in the Dipper's handle to the bright star Arcturus. This marks the buckle in the herdsman's Boots. Make a "spike" southward to the bright star Spica marking the spike of wheat held in the Lady's hand.

- *Leo the Lion*
- *The Chariot (also appearing as the Big Dipper and the Plow)*
- *The Herdsman's Boots*
- *Virgo's Spike of Wheat*

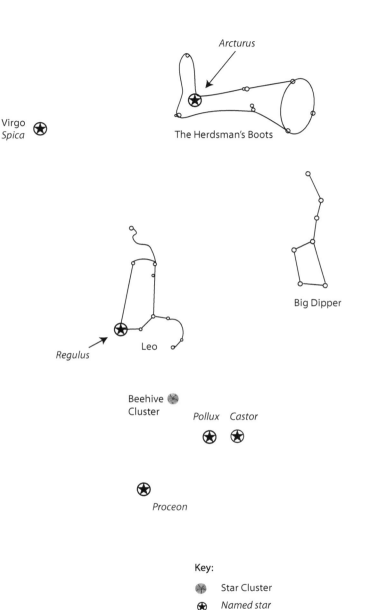

Spring Sky Map
Face west and look up

Leo the Lion
CONSTELLATION: LEO

In the old days, many important people had chariots to ride in. The richest people had the biggest and fanciest chariots, pulled by splendid horses. But someone had to be very special to have a chariot pulled by lions! Except for Dorothy's friend the Cowardly Lion in *The Wizard of Oz*, lions often represent great courage. Ours is a peaceful lion, resting in the shade of the lady's Chariot.

Regulus is the brightest star in Leo. It is named after a brave Roman general, captured in a war by the Carthaginians. They offered to let him go if he would tell the Roman Senate to give up the war. Regulus thought it was important for the Romans to keep fighting, so he went to Rome, but told the Senate not to give up. Then Regulus told his friends that he could not honorably accept his freedom since he had not kept his half of the bargain. He turned around and sailed back to Carthage to face certain death. His star in Leo is a symbol of his astonishing courage.

Star picture: Leo looks very much like a resting lion. The head is like a backwards question mark; the haunches are a triangle of stars.

Impress your friends

Stars rotate or spin, just as planets do. Our sun turns about once in 25 days. Regulus is a very fast spinner, rotating every 16 hours. This makes the star bulge in the middle, more like a flattened pumpkin than a round ball.

In binoculars

Regulus is the brightest star in Leo and is sometimes known as the Heart of the Lion, *Cor Leonis*. In the drawing below Regulus is in the lion's "elbow." If you focus on this light blue star, you might see a fainter yellow star up and slightly to the right. The yellow star is in orbit around Regulus, forming a double star.

Leo the Lion

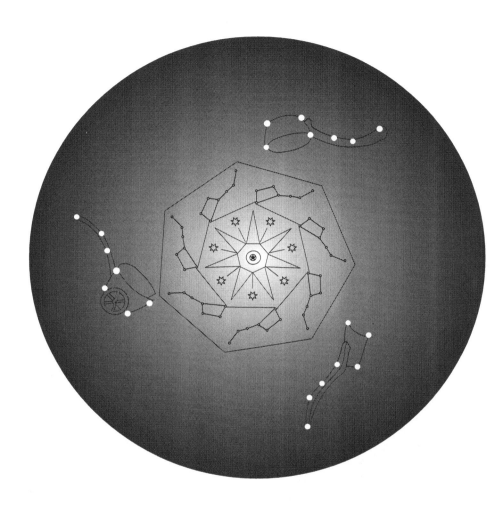

*The Nightly Dance Around Polaris
by the Plow, the Chariot, and the Dipper*

The Chariot, the Plow, the Dipper
Constellation: Ursa Major

What a versatile actor these stars are, playing three separate parts in our story. This is one of the most popular groups of stars in the night sky and has been since the earliest times. Some ancients saw the stars as a Chariot. The British see it as an old-fashioned Plow. Most Americans recognize it as the Big Dipper. Whatever the name, it is part of a very old constellation named Ursa Major, or the Great Mother Bear. But that's a story for another time.

The star picture is an arc of three stars connected to a square. A line through the arc helps you to arc to the star Arcturus. From there you can spike down to find the star Spica.

The front two stars of the Dipper, or the back two stars of the Chariot, are called the pointers. A line through these stars points to Polaris, the North Star. This information can be very useful if your chariot is lost, and you are trying to find directions on a dark starry night.

Impress your friends

Here's an ancient eye test for you. The middle star of the handle is really two stars. The bright one is called Mizar and the dimmer star is Alcor. Alcor was said to be a rider on Mizar, the middle horse who is pulling the plow or the chariot, whichever you prefer.

Most of the stars in the dipper are hot young stars, burning hydrogen. The top pointer star, named Dubhe, is much older. Dubhe has burned up its hydrogen and has started burning helium instead. It is 300 times more luminous than the sun.

In binoculars
Can't find little Alcor? It's easy in binoculars.

The Boots
CONSTELLATION: BOÖTES, THE HERDSMAN

A herdsman was an important fellow in ancient times. Tending animals such as cows, goats, pigs, and sheep was a lot easier than hunting for reindeer or woolly mammoths. People drank milk and ate cheese from cows or goats. They could knit sweaters, blankets, and socks from the wool of sheep. Pork, mutton, and beef were good to eat, and cowhide made good sandals and very nice boots. But herding was hard work.

One of the most dramatic changes in the history of the world was when people learned to plant and harvest wheat. Wheat could be ground and made into flour. Flour could be stored for the winter and used to make bread whenever people were hungry. Our herdsman wasn't the only fellow who decided to trade in his sandals for a nice pair of boots and become a wealthy wheat farmer.

In our star picture we see one tall boot, with a bright star at the buckle and dimmer stars for the heel, toe and top of the boot. These stars are part of the constellation traditionally called Boötes (Bo-oh-tees) the Herdsman. Boötes is easier to say in Greek than in English, so we just say Boots.

Impress your friends

The north pole of our galaxy is halfway between Leo and Boötes. When this point is overhead, the Milky Way can't be seen anywhere in the sky because it is circling all around the horizon.

Arcturus, the bright star in Boötes, isn't in the disk of our galaxy like most stars. It is orbiting up and down around the galactic center, instead of round and round in the galactic arms. Arcturus is the brightest star in the spring and summer skies.

The Boots

Virgo's Spike of Wheat
CONSTELLATION: VIRGO

Many cultures around the eastern Mediterranean Sea had a goddess they called either Our Lady or Mother Earth. This goddess had many names because each group said Lady or Mother in its own language. Often she was associated with the stars of the constellation Virgo and especially with the star Spica. Perhaps this is because Spica comes up in the spring when green plants reappear after a dreary winter. Maybe it reminded farmers to harvest their wheat when Spica sets with the sun on midsummer evenings.

Demeter is the Great Lady in our story. Her name means Mother Earth, and she was said to have the power that makes plants grow. When she was unhappy, there could be famine and drought. When she was in a good mood, there were fruits and grains in abundance. Demeter is said to have toured Greece in a chariot, teaching women how to grow wheat by hand. Men enjoyed eating bread, but thought wheat farming was women's work. When a herdsman helped Demeter find her daughter, she showed him how to make a large plow that could be pulled by an ox or a horse. With the ideas of heavy equipment and big animals pulling things, men could get excited about wheat farming, too.

The star picture is a single bright star, Spica. Virgo is a faint constellation, but well known because she is part of the Zodiac. Traditionally, Spica represented the head of a wheat stalk held in Virgo's hand. This head is often called an ear or spike of wheat.

Impress your friends

Spica is bright blue, the color of hot, fast-burning stars. It is more than two thousand times as luminous as our sun. Since the north pole of our galaxy is near Spica, there aren't many stars to see. This makes it a good direction for big telescopes to look beyond our Milky Way for other galaxies with billions of their own stars.

THE SUMMER SKY

Summer is a great time for star gazing. The evenings are often warm, and while summer doesn't have as many bright stars as the winter sky, it seems to have more stars overall. In fact there are ten star pictures in the summer sky, so we have to tell three stories to fit them all in. Summer is also a good time to watch out for snakes, and we'll find one or two of them hiding among our stars.

- *The three early summer constellations are high overhead, but faint.*

- *The four midsummer constellations are lower to the south, and brighter.*

- *The three late summer constellations are again high overhead, but bright and easy to see.*

- *Two spring constellations are visible through much of summer, the herdsman's Boots and the Big Dipper.*

Early Summer: Hercules and the Golden Apples

You may remember Cepheus, the king. His hobby was gardening, and his formal garden was very beautiful. At its center was an apple tree, on which grew the most wonderful golden apples. Lots of kings had red apples, but golden apples were rare and especially good for baking. Some people would try to steal a few when they thought no one was looking. Cepheus was normally a generous and kindly king, but he had a fondness for desserts. Cassia made a delicious apple pie . . . with cinnamon, of course. So Cepheus proclaimed his apples were "For Royal Use Only." He spread a rumor that his garden was protected by a dragon named Draco. That story kept most folks from daring to climb the wall.

The King of Tiryns wanted some of the golden apples, and he ordered a man named Hercules to get him some. Hercules was the greatest hero in all of Greece, and he wasn't afraid of anything. He set out on a dark night to climb the wall, and bring back the apples. He hoped to sneak past Draco, but the dragon's eyes could see in the dark. Draco roared and breathed a little fire. Hercules ran back to the wall and jumped over it in one great leap. Later Hercules got his friend Atlas to retrieve the apples for him, but that's a story for another time.

What to look for in the sky

First draw a line between the brightest stars Arcturus and Vega. North of the line is the diamond-shaped head of Draco the Dragon. South of the line is the square body of Hercules, looking a little like Orion, but fainter. Next, follow the pointer stars in the Plow to find Polaris, the pole star. Starting from Polaris, trace the long stem south to find the Apples dangling above Draco's coils.

- *The Golden Apples (portrayed by the Little Dipper)*

- *Draco the Dragon*

- *Hercules*

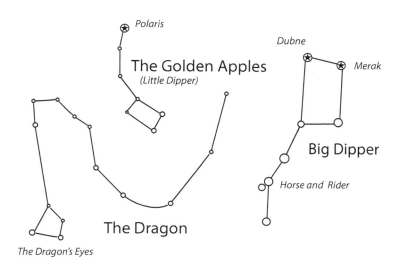
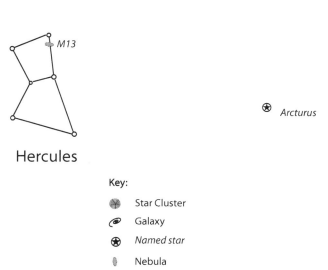

Key:

- Star Cluster
- Galaxy
- Named star
- Nebula

Early Summer Star Map
Face south and look up

The Golden Apples
Constellation: Ursa Minor, Little Bear (Little Dipper)

Apple gardens were special places in ancient times. King Cepheus' garden was a splendid example. At the entrance was a magnificent palm court and the palace where Queen Cassia conducted affairs of state. Inside the garden, besides the golden apple tree, were many other exotic plants and animals from far off places. Later in summer, you'll notice some autumn stars rising, including those of Cepheus' Dark Tower. Maybe he's up early to watch over his apples.

The star picture is a pair of apples hanging from Polaris, the North Star. This constellation is commonly known on star maps as Ursa Minor, the Little Bear. Another name for it is the Little Dipper. Near city lights several of the Little Dipper's stars are too faint to see, but two of them are bright enough to play the apples in our story. To find Polaris, recall that you can follow the two pointer stars in the Big Dipper.

Impress your friends

The star Polaris marks the North Pole in our view of the sky. It sits almost directly above Earth's North Pole, so it doesn't seem to move as Earth spins below it. Nearby stars and constellations rotate around the pole, but never set below our horizon. Some of these "eternal" constellations are Ursa Major, Ursa Minor, Cepheus, and Cassiopeia.

Polaris is the closest and brightest cepheid variable star. Cepheid variables are yellow super-giants, and Polaris is 2,000 times more luminous than the sun.

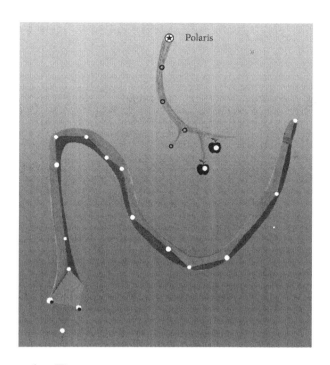

Draco the Dragon
Constellation: Draco

Tales about ancient gardens had several things in common. They all had walls and apple trees, and each was guarded by a dragon or a snake. Often the dragon's name was L'Adon, which meant "the lord of the garden." Back before cats became popular people kept a pet snake to chase the mice and rats away from the kitchen pantry and from their gardens. So maybe Draco was a pet garden snake, and maybe Cepheus nicknamed him lord of the garden to scare the mice. In any case, Draco rarely shared out the apples to just anyone. If you don't believe me, ask Hercules!

The star picture: The Dragon is faint, but actually bears a good resemblance. To find Draco, look for his bright eyes and diamond shaped head west of the star Vega, about one-third the way to Arcturus, and a little south of the Apples. His body snakes around the apples, guarding them from Hercules.

Hercules
Constellation: Hercules

Hercules is trying to hide from Draco the Dragon, so he can steal some of Cepheus' golden apples. He's hiding so well, that sometimes he's hard to find! If you can spot the eyes of the dragon, they are looking straight at Hercules.

The star picture is the body of Hercules. Remember that he went to the garden on a very dark night. The stars of Hercules are so faint, you might need a dark, moonless night to spot him. He's also located along a line between the two brightest summer stars, Arcturus and Vega. Many star maps show Hercules upside down, but we see him right side up looking for the Dragon.

In binoculars
M13, the Hercules cluster, is a beautiful **globular cluster**. Globular clusters are old and rare - only about 150 are known. They are much bigger than open clusters. A large open cluster might have one hundred stars. The Hercules globular cluster has hundreds of thousands.

*Hercules on the lookout for Draco
Do you think he'll find him?*

MIDSUMMER: THE THREE KINGS

Our next group of constellations includes three members of the zodiac: Libra the Scales, Scorpius the Scorpion, and Sagittarius the Archer. There is no traditional story linking these characters, so we'll borrow a story from the Greek historian Herodotus about three ancient kings and the mysterious Oracle of Delphi.

Croesus of Lydia was a very wealthy king with a treasury full of silver, gold, and precious jewels. People say he invented money, minting the first gold coins ever made. One day he was visited by Solon, the *archon* of Athens. (An archon is a sort of elected king.) Solon was famous as a wise man and a just lawgiver. Croesus gave a banquet in his honor and asked Solon the name of the happiest man he had ever met. Of course Croesus thought Solon would say that Croesus was the happiest. Instead Solon mentioned several regular people who had lived happy lives and had good families. Croesus protested, "What about me? I have so much more wealth and power than they had." Solon wisely said that Croesus was happy now, but his life wasn't over yet, so it was impossible to say whether or not he would always be happy.

Later Croesus decided to attack his neighbors the Persians. But first he wanted to ask the Oracle of Delphi whether he would win or not. He sent great treasures to the Oracle, including a beautiful golden brooch in the shape of a scorpion, with a bright red ruby at its heart. The Oracle said, "If you attack Persia you will destroy a great kingdom." Croesus was thrilled with this message and immediately sent out his army. Unfortunately for him, Cyrus the Great was king of Persia at the time, a good general, and had the famous Persian archers in his army. He defeated Croesus and captured his capital city of Sardis. Croesus had destroyed a great kingdom, but instead of Persia it was his own. With his city in flames Croesus was heard to cry out, "Oh wise Solon, you were right. I am the most miserable of men."

What to look for in the sky

The symbols for the three kings are found toward the southern horizon. As part of the Zodiac they mark the ecliptic, the sun's yearly path through the stars. Look first for a very bright red star marking the heart of the Scorpion. The beautifully jeweled Scorpion represents Croesus. Trace out its short claws and long curving tail. In front of the Scorpion look for two medium stars, one above the other. These mark Libra, the scales of justice, but we'll think of it as the staff of Solon the lawgiver. Behind the Scorpion is a broad group of stars forming Sagittarius the Archer, representing King Cyrus of Persia. Finally, above these three star pictures, look for the curving outline of a serpent. The symbol of the Oracle of Delphi was a serpent called Python.

- *The Staff of Solon (Libra)*
- *The Scorpion*
- *The Persian Archer (Sagittarius)*
- *The Python of Delphi (Serpens)*

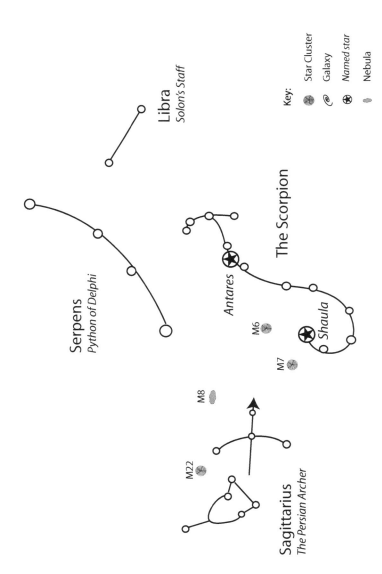

The Staff of Solon
Constellation: Libra the Scales

Solon, the great lawgiver of Athens, was hailed as one of the Seven Wise Men of antiquity. His staff or walking stick was his symbol of justice. The people of Athens elected him archon, and asked him to give them a new set of laws. Solon worried that the people would reject the new laws. He told them he had to go on a trip and made them promise not to change the laws until he got back. He traveled for several years, and by the time he returned the Athenians were used to the new laws and didn't want to change them any more. It was on this long journey away that Solon met Croesus in the city of Sardis. The official constellation is called Libra, the Scales.

The star picture: Most of the stars that make up Libra are faint. Only two are bright enough to show near city lights. Look for these two stars representing Solon's Staff between the star Spica in Virgo and the claws of the Scorpion.

Impress your friends

Antares, the heart of the Scorpion, is a red super-giant star. The only other red super-giant as bright as Antares is in the winter sky. Remember Betelgeuse in the constellation Orion? These two stars are each 500 light years away in opposite directions, so we are about halfway between them. Antares is a very young star and much heavier than our sun. The bigger stars are, the faster they burn out. Our sun will burn for billions of years. But Antares is nearing the end of its life already, even though it started burning only 20 million years ago.

The Scorpion
CONSTELLATION: SCORPIUS

We think of the Scorpion as a golden brooch from the treasures of King Croesus, who misunderstood the advice of the Oracle and started a war with King Cyrus of Persia. The Scorpion is facing Solon's Staff, and trying to sting Cyrus' Archer.

The star picture: Scorpius is one of the best constellations because it actually looks like a scorpion studded with jewels, including the ruby-red star Antares at its heart. The curving tail is marked by a bright diamond-like star called Shaula, which is Arabic for *the sting*.

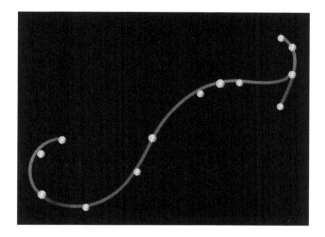

In binoculars
Shaula is the bright blue star at the tip of the scorpion's tail. Above and to the east of Shaula, you will find two open star clusters named M6 and M7. Each has about 100 stars.

M6 is called the Butterfly Cluster because in a telescope it looks like a butterfly flying toward the east.

The Persian Archer
CONSTELLATION: SAGITTARIUS

Ancient Persian armies were famous for their archers. In fact some Persian gold coins had a picture of an archer on them. Sagittarius is often drawn as a centaur, a mythical creature that is half man and half horse. This can be a bit confusing since there is another totally different constellation called *The Centaur* just below the horizon.

The star picture is a Persian soldier with a bow and arrow. Look for the 3 stars of the bow and another star marking the arrowhead, aiming at the Scorpion, then find the archer's head with his tall Persian hat. Some people think Sagittarius resembles a teapot, with the spout where the archer's bow and arrow are. The bright region of the Milky Way that looks like steam coming out of the spout of the teapot is the galactic center, about 30,000 light years away.

Impress your friends

The center of our Milky Way galaxy lies in Sagittarius. A large black hole was found there, and is called Sagittarius *A*.

In binoculars

Because Sagittarius marks the center of our galaxy there are many wonderful things to see. If you look at the arrow through binoculars then slowly scan up from there, you will come across the Lagoon Nebula (M8). This is a cloud of dust and hydrogen gas where stars are born. M22 is a globular cluster with about 75,000 stars. It is among the closest, appearing larger in binoculars than M13 in Hercules.

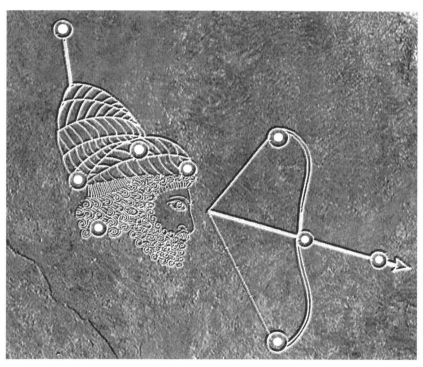

Sagittarius the Archer

The Python of Delphi
CONSTELLATION: SERPENS

In ancient times if you wanted to know the future you could go to a special place and ask the Oracle there to tell you. There were several such places in Greece, but the most famous was Delphi. Its symbol was a serpent named Python, and the priestess who gave advice and predicted the future was called the Pythoness. Her advice wasn't always very clear and could easily be misinterpreted. That's what happened to Croesus when she told him "You will destroy a great kingdom."

The star picture: Over the heads of the Scorpion and the Archer you'll find a reasonably good, curvy serpent. The stars aren't bright, but they can usually be made out.

The Serpent used to be part of a very big constellation, showing an unnamed hero wresting with a giant snake. Today there are three constellations, the Serpent's Head, the Serpent's Tail, and Ophiuchus, which means the Serpent Bearer. Above the Serpent, in the direction of Hercules, you might make out an outline of the head and shoulders of Ophiuchus.

Oddly enough, there is a legend about the Oracle of Delphi in which the god Apollo wrestled the Python in order to gain control of the Oracle for himself. Ophiuchus is never given a name, but perhaps he was Apollo.

LATE SUMMER: THE SUMMER TRIANGLE: THE LYRE, THE SWAN, AND THE EAGLE

The Summer Triangle appears late in the Summer, and stays in the evening sky through the autumn and well into December. It is one of the easiest star pictures to find, and it is formed by a bright star from each of three constellations: the Lyre, the Swan, and the Eagle. There are unrelated stories about each of these constellations, so we're going to tell a new story to bring the three together.

The most famous musician of ancient Greece was Orpheus. He played an instrument like a harp, called a lyre, while he sang his poems. His music was so beautiful that even the gods would grant him favors to play for them. He sang about love and beauty, and also about the adventures of mythical heroes. Perhaps he sang about Daedalus, the first man who learned how to fly.

Daedalus was an engineering genius who worked for King Minos. He designed a beautiful palace for Minos, complete with hot and cold running water. He also designed a labyrinth in which to confine the dreaded Minotaur, a monster who was half man and half bull. Daedalus wanted to travel and see the world, but Minos refused to let him leave his kingdom, confining him to the palace grounds. Daedalus set to work building winged gliders, one for himself and one for his son Icarus. His was made of large, white swan feathers to carry his weight. For his son's glider he used more aerodynamic eagle feathers. Early one morning, when the gliders were ready, they stood on the palace wall above a high cliff and launched themselves into the air. The wings worked! Off they flew. Icarus was so excited, he wanted to fly higher and faster than his father. He flew so high that the wax holding the feathers onto his glider began to melt from the heat of the sun. His glider tipped, and Icarus fell down, down into the sea.

What to look for in the sky

Vega rises in the east in midsummer, and the whole triangle is visible by late summer, high overhead. The brightest of the three points of the Summer Triangle is Vega, in Lyra. The bright star furthest south is Altair, the head of the Eagle. The bright star west of Vega is Deneb, the tail of the Swan.

- *The Lyre*

- *The Eagle*

- *The Swan*

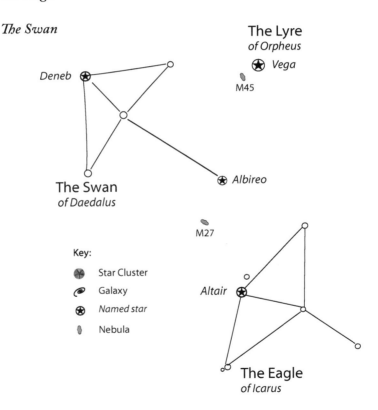

Late Summer Star Map

Face south and look up

The Lyre of Orpheus
CONSTELLATION: LYRA

Orpheus was the most famous mythological Greek musician. He sang beautiful poems while playing on the lyre, a stringed instrument like a harp given to him by his father Apollo. When Orpheus' wife Eurydice died from a snake bite he traveled to the underworld. He hoped to rescue her by enchanting Hades, the king of the underworld, with his music. Hades agreed to release Eurydice on condition that Orpheus not look back until they were out of the underworld. They almost made it, but near the exit Orpheus glanced back to make sure Eurydice was still with him, and she disappeared.

The star picture is a small, faint rectangle including the bright star Vega. On most nights it is a one-star lyre, featuring only Vega.

Impress your friends

Beta Lyrae is an eclipsing binary star, like Algol in Perseus. Beta Lyrae's two stars orbit each other every 13 days. When one of the stars passes in front of the other every 6½ days, it dims to half its brightness, because we only see the light from one of the stars.

The sun and our solar system are circling in the Milky Way. We are traveling toward the star Vega, getting closer at the rate of a million miles a day.

In binoculars
Lyra is home to M57 the famous Ring Nebula. This is a planetary nebula±a shell of gas blown off by a dying star lighted by radiation from the inner star, now a white dwarf. The Ring is very small and so faint you'll need a 4-inch telescope.

The Swan Hang Glider
Constellation: Cygnus the Swan

To bring Cygnus the Swan into our story we turned him into a winged glider built by Daedalus. Daedalus is trying to escape. To fool king Minos his swan-kite is flying backwards as he flees toward the tower of his friend King Cepheus.

The star picture: The bright star Deneb marks the tail of the Swan. Deneb is Arabic for *tail*. Reaching out in a triangle are the stubby wings. Finally look for one lonely star way out in front, marking the beak of the swan. This star is called Albireo. In our star picture, we see a giant glider with wings made of swan's feathers. Daedalus is marked by Deneb as he guides the winged kite through the sky.

In binoculars

In a 4-inch telescope the star Albireo is a beautiful double star, one yellow and one blue. However, in binoculars it usually looks like a single star.

Impress your friends

Deneb is one of the largest super-giant stars in our galaxy. It looks about as bright as Altair, the bright star in the Eagle, but Altair is only 16 light years away, and Deneb is 1,500 light years away, almost 100 times farther. Altair is 10 times more luminous than the sun, and Deneb is 50,000 times more luminous than the sun. If Deneb were as close as Altair, it would be visible in broad daylight and would make shadows at night.

Daedalus Riding his Swan Glider

The Eagle Hang Glider
Constellation: Aquila the Eagle

In our story Aquila the Eagle is another winged glider. This one was built by Daedalus for his son Icarus. Eagle feathers must have produced a faster, high-performance glider. Icarus flew too high and too fast for his own good, and we see him falling off and into the great southern sea.

The star picture: The bright star Altair marks the head of the Eagle. Three slightly dimmer stars mark the tip of each wing and the tail. Icarus is marked by the star near Altair, falling from his winged-kite into the sea.

Impress your friends

Eta Aquilae, near the left wing tip of Aquila, is one of the four easily visible Cepheid variable stars. It has a brightness cycle of about 7 days.

Altair is another fast-spinning star like Regulus in Leo. Altair spins even faster, rotating once every 6 1/2 hours.

In binoculars

Between the Eagle and the Swan is M27, the Dumbbell Nebula. It is another planetary nebula like the Ring Nebula in the Lyre. Fortunately it is closer to us, and therefore brighter, so you have a better chance of finding it in your binoculars.

Icarus Falls from his Eagle Glider

STARS AND PLANETS

In this book you've learned to use star pictures to locate a number of constellations. You've also learned the names of 20 of the brightest stars. As you may already know, a star is a huge burning ball of hydrogen or helium gas. Its energy comes from nuclear reactions among gas particles, giving off incredible amounts of light and heat. You can tell how hot a star is by its color. Blue and white stars are much hotter than red ones.

Stars have two kinds of brightness. Luminosity, sometimes called absolute brightness, measures how bright a star is up close. Apparent brightness indicates how bright it looks to us, here on Earth. Stars that look bright to us are either fairly close like Sirius and Procyon, or extremely luminous like Rigel, Betelgeuse, and Deneb. All the bright stars we see are larger and brighter than our sun. The most luminous stars are hundreds or even thousands of times more luminous.

We measure astronomical distance using the term *light years*, the distance a beam of light travels in one year. Our sun is about 8 light minutes from Earth. Sirius, the closest bright star, is about 8 light years away. That's 525,600 times further than the sun. Several stars are closer to us than Sirius, but they are much fainter. Alpha Centauri is the closest visible star at 4.3 light years. It is about the same size as the sun. Proxima Centauri is slightly closer, but too dim to see.

Our Milky Way galaxy is about 100,000 light years across. It has a center and several spiral arms. Our solar system is about 30,000 light years from the galactic center. All the individual stars you can see without using binoculars or a telescope are fairly close to us. In fact they are in the same arm of our Milky Way galaxy that we are, the Orion Arm. Deneb, the farthest bright star on our list, is only 1,500 light years away. We are orbiting the Milky Way once every 250 million years. Each orbit is called a galactic year. The sun has made about 20 orbits so far in its lifetime.

If you see a bright star where there shouldn't be one, it's probably a planet. Of the eight planets, there are only four that you're likely to see in the sky. Venus looks much brighter than any star, but being close to the sun never appears overhead. It may appear in the east before dawn or the west after sunset. Jupiter is also brighter than any star, almost as bright as Venus. It is farther from the sun, so it may appear near the horizon or high overhead. Saturn and Mars are the trickiest, looking about as bright as many of our bright stars. If it is reddish, it's Mars. A good clue is that planets are always near a constellation of the Zodiac. So you might find Saturn or Venus near Leo or Taurus, but never near Orion or Cassiopeia.

In ancient times, many people thought the Milky Way was a distant cloud. Democritus, a Greek scientist living around 400 BCE, suggested it was simply the light of millions of distant stars. Two thousand years later, Galileo was the first person to see these distant stars, using a telescope. Lucretius, a Roman scientist living about 60 BCE, speculated that there were "other Earths" near other stars, and that some of those planets would have people and animals living on them. Two thousand years later we are seeing the first satellite pictures of the planets he predicted.

Turning on the Stars

There's a game you can play in the winter or the summer, called "turning on the stars." Stars begin to come out shortly after sunset. Which star will come out first? The brightest one of course, and then the second brightest, and so on. The table of bright stars on the next two pages is organized by season and by brightness. You can use this list to predict the order that stars will appear. With a little practice you can make it look as if you are turning on the stars one at a time.

You'll need to practice so you'll know where the stars are going to be before they appear. Then you can point in that direction and say, for example, "Our first star tonight will be Sirius." Next point high in the sky and say "Next how about Capella." A few seconds later you can add "Now for Rigel," while you point a little above Sirius. Try it a couple nights in a row to get your timing right.

For the summer there are two things you'll want to know. Although Arcturus is a spring star, it stays visible for much of the summer. Spica is also visible into early summer, so you could add these two stars to your show. Arcturus is slightly brighter than Vega, so it should appear first. But later in the summer, as it gets closer to the setting sun, twilight will hide it for a little while, so Vega will appear first. There's also a nice finishing touch for summer. Right after the last bright stars Deneb and Shaula, the entire handle of the Big Dipper appears all at once. This is because the three handle stars are equally bright.

One last tip: young people can often see a star a minute or so before grown-ups, because their eyes are sharper. So if a mom or dad is turning on the stars for some youngsters, be prepared for the kids to see the stars first.

List of Bright Stars

These are the 20 brightest stars in the constellations of our stories. Winter has the most bright stars. Summer has the most visible stars including several bright ones, and spring has the fewest visible stars. Autumn has plenty of visible stars, but no bright stars at all.

Winter Stars	Constellation	Brightness Rank	Distance* (Light Years)	Comments
Sirius	Canis Major	1	8	Closest bright star; eye of Big Dog
Capella	Auriga	4	45	Auriga's "cap"
Rigel	Orion	5	800	White giant 3rd furthest away
Procyon	Canis Minor	6	10	2nd closest bright star
Betelgeuse	Orion	7	470	Red giant 4th furthest away
Aldebaran	Taurus	9	60	Red eye of the bull
Pollux	Gemini	12	40	The brighter twin
Adhara	Canis Major	15	570	
Castor	Gemini	16	50	
Belatrix	Orion	18	240	Orion's dimmer shoulder
El Nath	Taurus	19	130	The tip of the bull's horn
Alnilam	Orion	20	1,340	2nd furthest away Middle star in Orion's belt

*Distances are approximate.

Spring Stars	Constellation	Brightness Rank	Distance* (Light Years)	Comments
Arcturus	Boötes	2	35	2nd brightest
Spica	Virgo	10	260	
Regulus	Leo	14	70	

Summer Stars	Constellation	Brightness Rank	Distance* (Light Years)	Comments
Vega	Lyra	3	25	4th closest 3rd brightest
Altair	Aquila	8	18	3rd closest
Antares	Scorpius	11	365	Red giant
Deneb	Cygnus	13	1,500	Furthest away
Shaula	Scorpius	17	330	

Autumn Stars	Constellation	Brightness Rank	Distance* (Light Years)	Comments
No bright stars				

*Distances are approximate.

Constellations of the Zodiac

The path the sun, moon, and planets follow through the sky is called the ecliptic. Traditionally the 12 constellations that lie along this path are called the Zodiac. These are the time of year they can best be seen in the evening sky.

Winter: Taurus the Bull, Gemini the Twins, Cancer the Crab*

Spring: Leo the Lion, Virgo the Lady, Libra the Scales

Summer: Scorpius, Sagittarius the Archer, Capricorn the Sea Goat*

Autumn: Aquarius the Water Carrier*, Pisces the Fish*, Aries the Ram

* <u>Note:</u>
These 4 "watery" constellations are too dim to see in suburban skies.

GLOSSARY

Asterism: A pattern of stars. Some asterisms are part of a larger constellation, such as Orion's Belt in Orion, or the Big Dipper, which is part of the Great Bear. Other asterisms include stars from several constellations. The Summer Triangle includes one star from each of three constellations: the Lyre, the Swan, and the Eagle.

Binary star: Two or more stars connected by gravity, and orbiting a central point. Castor in the Twins is a binary star.

Black hole: Region of space whose gravitational field is so strong nothing, even light, can escape: for example Sagittarius A.

Cepheid variable: A single star that becomes brighter and dimmer over a regular period related to its average brightness. This relationship allows astronomers to easily calculate its distance. The first one discovered was delta-Cepheii in Cepheus.

Constellation: Traditionally, one of 44 star patterns listed by ancient Greek astronomers Eratosthenes and Ptolemy of Alexandria. Modern astronomers have redefined the term, dividing the sky into 88 regions, each named as a constellation. The stars of each classic constellation are included in one of the modern constellations, along with any extra stars in its vicinity.

Double star: One that appears to be a single star, but when seen through a telescope or binocular is actually two stars. They may or may not be orbiting one another. Albireo in the Swan is a double star.

Eclipsing variable: A binary star system aligned just right, so that the stars pass in front of one another on every orbit as seen from Earth. The stars appear to "wink" as they periodically block each other's light. Algol in Perseus is an eclipsing variable star.

Ecliptic: The yearly path of the sun mapped across the sky in a great circle. See "Zodiac."

Nebula: A luminous gas cloud. One type, called an "emission nebula," shines by re-radiating light absorbed from the light of the stars within the cloud. Others shine by simply reflecting the stars' light. M42 in Orion is an emission nebula. See also "Planetary nebula."

Galactic cluster: See Open cluster.

Galaxy: A galaxy is sometimes called an "island universe." Each galaxy consists of billions of stars, gravitationally bound into a spiral, spherical, or elliptical shape. Galaxies are separated from each other by incredible distances of empty intergalactic space, and there are over 100 billion of them. Our solar system is part of the Milky Way galaxy, along with all the stars, clusters, and nebulae we can see in the night sky. M31 in Andromeda is our largest neighboring galaxy.

Globular cluster: A very dense cluster of stars shaped like a globe, holding tens of thousands of stars. There are only about 120 of them. They are very old, and found near the center of the galaxy. M13 in Hercules is a globular cluster.

Open cluster: An unstructured group of stars, having a few dozen or a few hundred stars. These are primarily new stars born from a dense gas cloud. There are about 700 known open clusters. The Pleiades is an open cluster.

Planetary nebula: A small, circular glowing object that looks like a planet. It is actually a shell of gas blown off from a collapsing star. They are very hard to see in small telescopes. M57, the Ring Nebula in Lyra, is a planetary nebula.

Star picture: See Asterism.

Zodiac: "The circle of the animals." A ribbon of sky a few degrees on either side of the ecliptic, and especially the 12 traditional constellations that lie along this ribbon. The Zodiac marks the portion of the sky traveled by the sun, the moon, and planets. Leo the Lion is a constellation in the zodiac.

INDEX

A

Aegenor 1, 9
Agamemnon 26
Andromeda galaxy 7, 9, 14
Apollo 52
asterism 67
   Andromeda's Chain 1ff, 9
   Aries, the two-star sheep 11, 25
   Big Dipper 27ff, 33
   Cassia's Crown 1ff, 4
   Cepheus' Tower 1ff, 7
   Chariot 27ff, 33
   Chariot Wheel 17ff, 23
   Eagle Hang Glider 53ff, 58
   Eyes of the Dragon 38ff, 41
   Golden Apples 38ff, 40
   Hercules 38ff, 42
   Herdsman's Boots 27ff, 34
   Horse and Rider 33, 39
   Leo the Lion 27ff, 30
   Little Dipper 40
   Lyre of Orpheus 53ff, 55
   Major The Big Dog 17ff, 24
   Minor The Little Dog 17ff, 25
   Orion's Belt 17ff, 20
   Pegasus, Great Square of 1ff, 10
   Perseus' Shield 1ff, 12
   Persian Archer 45ff, 50
   Pleiades 14
   Plow 27ff, 33
   Python of Delphi 46, 52
   Scorpion 45ff, 49
   Solon's two-star staff 45ff, 48
   Swan Hang Glider 53ff, 56
   Taurus the Bull 17ff, 22
   Teapot See Sagittarius
   The Twins 17ff, 26
   Virgo's Spike of Wheat 27ff, 36

B

binary star 67
black hole 50, 67

C

Carthage 30
Centaur 50
Cepheid variable stars 7, 58, 67
   Delta Cephei 7
   Eta Aquilae 58
   Polaris 40
   Zeta Geminorum 26
Clytemnestra 26
constellations
   Andromeda the Princess 9
   Aquila the Eagle 58
   Aries the Ram 11, 66
   Auriga the Chariot Driver 23
   Boötes the Herdsman 34
   Canis Major the Big Dog 24
   Canis Minor the Little Dog 25
   Cassiopeia the Queen 4
   Cepheus the King 7
   Cygnus the Swan 56
   Draco the Dragon 41
   Gemini the Twins 26, 66
   Hercules the Hero 42
   Leo the Lion 30, 66
   Libra the Scales 48, 66
   Lyra the Lyre 55
   Orion the Hunter 20
   Ophiuchus the Serpent Handler 52
   Pegasus the Flying Horse 10
   Perseus the Hero 12
   Sagittarius the Archer 50, 66
   Scorpius the Scorpion 49, 66
   Serpens the Serpent 52
   Taurus the Bull 22, 66
   Ursa Major the Great Bear 33
   Ursa Minor the Little Bear 40
   Virgo the Great Lady 36, 66
Crete 1, 7
Croesus 45, 46, 48, 49, 52
Cyrus 45, 46, 49

D

Daedalus 53ff
Delphi 45, 46, 52
Demeter 17, 27, 36
Democritus 62
Demon Star 12
Dog Star 18, 24
double star, definition 67

E

eclipsing binary star
  Algol 12, 55
  Beta Lyrae 55
  definition 67
ecliptic 11, 22, 46, 62
  definition 67
emission nebula, definition 67
Eurydice 55

G

galactic center 34, 50, 61
galactic star cluster, definition 68
galactic year 61
galaxy, definition 68
  Orion Arm of the galaxy 61
Galileo 62
globular star cluster, definition 68
Gorgon 1, 2, 9, 12ff
Greece 27, 36, 38, 52, 53

H

Hades 55
Helen of Troy 26
Herodotus 45
Horse and Rider 33, 39
Hubble, Edwin 7

I

Icarus 53, 58, 59

J

Jaffa 1, 4, 7, 10

L

L'Adon 41
labyrinth 53
Leavitt, Henrietta 7
Lucretius 62
Lydia 45

M

Messier, Charles 14
Messier objects
  M6, Butterfly Cluster 49
  M7 49
  M8, Lagoon Nebula 50
  M13, Hercules Cluster 42, 50
  M22 50
  M27, Dumbbell Nebula 58
  M31, Andromeda Galaxy 9
  M41 25
  M42, Orion Nebula 21
  M44, Beehive Cluster 26
  M57, Ring Nebula 55
maps. See Star Maps
Mediterranean Sea 14, 36
Milky Way 4, 9, 10, 11, 20, 34, 36, 50, 55, 61, 62
  Orion Arm 61
Minos 53, 56
Minotaur 53
Mother Earth 27, 36

N

nebula
  definition 68
  Dumbbell Nebula 58
  Lagoon Nebula 50
  Orion Nebula 21
  Ring Nebula 55
North Star 7, 33, 40

O

open star cluster, definition 68
Oracle of Delphi 45, 46, 49, 52
Orion Arm of the galaxy 61
Orpheus 53, 55

P

Persephone 27
Persia 45, 50
planetary nebula, definition 68
planets
   Jupiter 22, 62
   Mars 20, 22, 62
   Saturn 22, 62
   Venus 22, 62
Pleiades 14
Pleione 14
Pluto 27
Python 46, 52
Pythoness 52

R

Regulus, Roman general

S

Sagittarius A 50, 67
Sardis 45, 48
Seven Sisters 14
Seven Wise Men 48
solar system 9, 11, 55, 61
Solon 45, 46, 48, 49
Sparta 17, 18, 26
star cluster names
   Beehive Cluster 26
   Butterfly Cluster 49
   Hercules Cluster 42
   Hyades Cluster 23
   Perseus Double Cluster 14
   Pleiades Cluster 1, 2, 12, 14, 23, 26
Star Maps
   about x
   Autumn 3
   Early Summer 39
   Late Summer 54
   Midsummer 37, 45, 47, 54
   Spring 29
   Winter 19
star names
   Adhara 19, 64
   Albireo 54, 56
   Alcor 33
   Aldebaran 19, 22, 64
   Algol 2, 3, 12, 13, 14, 15, 55
   Alnilam 19, 64
   Alpha Centauri 24, 61
   Altair 54, 56, 58, 65
   Antares 20, 47, 48, 49, 65
   Arcturus 28, 29, 33, 34, 38, 39, 41, 42, 63, 65
   Belatrix 19, 21, 64
   Beta Lyrae 55
   Betelgeuse 20, 21, 48, 61, 64
   Capella 3, 16, 18, 19, 23, 63, 64
   Castor 19, 26, 29, 64
   Cor Leonis 31
   Delta Cephei 7
   Deneb 2, 3, 7, 54, 56, 61, 63, 65
   Dubhe 33, 39
   El Nath 19, 22
   Eta Aquilae 58
   Merak 39
   Mizar 33
   Polaris 3, 7, 29, 32, 33, 38, 39, 40, 43
   Pollux 19, 26, 29, 64
   Procyon 18, 19, 25, 29, 61, 64
   Proxima Centauri 61
   Regulus 29, 30, 31, 65
   Rigel 19, 21, 64
   Shaula 49, 63, 65
   Sirius 18, 20, 24, 61, 63, 64
   Spica 28, 29, 33, 36, 48, 49, 65
   Vega 38, 39, 41, 42, 43, 54, 55, 56, 63, 65
stars
   apparent brightness of 61
   binary 67
   list of bright 64-65
   luminosity of 61
   turning on 63
   what are they? 61

star picture. See also asterism
   definition 68
super-giant stars 20, 48, 56

T

Tiryns 38
Triptolemus 27

W

wheat 27, 28, 34, 36

Z

Zodiac 11, 22, 26, 36, 45, 46, 62
   constellations of 66
   definition 68

ROB DREW

Rob's interest in astronomy began at age 7 with a book given him by his uncle. Nurtured with visits to the Griffith Observatory and Planetarium in Los Angeles, California, and the Boy Scout merit badge program, he eventually studied physics and astronomy at  Caltech in Pasadena. Although familiar with the details of technical astronomy, his main enjoyment remains star gazing and storytelling under a clear evening sky. His first attempts to point out constellations to his children fell short of success because, except for the Big Dipper, most constellations don't really look like they are supposed to. This led Rob to develop the system of star pictures and stories presented in this book. He has since taught classes and led star parties introducing students, friends, and neighbors to Cassia's crown, Perseus' shield, and the Eyes of the Dragon. Most recently Rob has entertained nighttime visitors to Oakland's Chabot Space and Science Center by helping them locate favorite constellations and telling their stories.

ELAINE DREW

 Elaine Drew became interested in art while working as a fashion designer in Manhattan. She works primarily on mythological subject matter in the medieval painting medium of egg tempera, an ancient technique that combines pure pigment with egg yolk. She served on the Board of Directors of the Society of Tempera Painters for seven years. Her work was featured on the cover of the 2008 winter issue of *Dream Time* magazine, published by the International Association for the Study of Dreams, and in the *Adobe Illustrator CS5 Wow! Book*. She is a member of Phi Beta Kappa.

Made in the USA
Lexington, KY
30 July 2015